Henry F. Noyes

How to Apply for a Patent

An explanation of the method of drawing up and prosecuting patent

applications with examples taken from actual practice

Henry F. Noyes

How to Apply for a Patent
*An explanation of the method of drawing up and prosecuting patent applications
with examples taken from actual practice*

ISBN/EAN: 9783337370602

Printed in Europe, USA, Canada, Australia, Japan

Cover: Foto ©berggeist007 / pixelio.de

More available books at **www.hansebooks.com**

How To Apply For A Patent

AN EXPLANATION OF THE
METHOD OF DRAWING UP
AND PROSECUTING . . . . .
PATENT APPLICATIONS
WITH EXAMPLES TAKEN
FROM ACTUAL PRACTICE

By HENRY F. NOYES, PATENT SOLICITOR

CONTENTS.

INDEX TO FORMS.

PREFACE.

The purpose of this book is to make clear the details of the practice of the United States Patent Office in patent applications; to the end that many of those who are not able to procure the services of experienced attorneys may still be able to protect their ideas.

The matter presented has been gathered from the "Rules of Practice" of the Patent Office, Decisions of the Commissioner of Patents and the United States Courts, several works on Patents, and from actual practice; the aim being to make the application of these rules, and the method of proceedure, so plain that the services of an attorney or agent will not be necessary.

Little reference has been made to the practice in interference, appeals, etc., since only a very small percentage of the applications ever develope into these disputes, and it would not be possible to cover the practice in this little work ; and it would not be advisable for anyone to attempt such cases without experienced technical assistance.

If this publication helps anyone to protect his attempts to promote the progress of the useful arts, its appearance will be justified.

HENRY F. NOYES.

SYNDICATE BUILDING,
NEW YORK, 1897.

1. ATTORNEY.

As there is no limitation prescribed by the Patent Office as to who shall act as attorney in the prosecution of a case, the papers may be prepared by the inventor, or by any one he shall appoint. As the value of the protection which a patent affords depends upon the skill and care with which the papers are drawn up, they should not be attempted until the inventor thoroughly understands the methods and practice. The Office practice requires regular forms to be followed in all cases. (See appendix.)

If an attorney be appointed, he has no power to act until his power of attorney shall have been filed, which is customarily done with the other papers relating to the case. (See appendix, form 3.)

The power of attorney may be revoked at any time. (See appendix, form 24). For gross misconduct the Commissioner may refuse to recognize any person as a patent agent, either generally or in each particular case.

2. WHAT MAY BE PATENTED.

Section 4886, Revised Statutes of the United States, says : " A patent may be obtained by any person who has invented or discovered any new and useful art, machine, manufacture or composition of matter, or any new and useful improvement thereof, not known or used by others in this country, and not patented or described in any printed publication in this or any foreign country, before his invention or discovery thereof, and not in public use or on sale in the United States for more than two years prior to this application, unless the same is proved to have been abandoned."

The three essentials to a patent thus are, invention, novelty, and utility.

The interpretation of the word "novelty" does not leave much chance for argument. The invention is either new or it is not. If not the records will tell. The question of novelty is a question of fact. Novelty is not, however, precluded by an abandoned application for the same thing, nor an abandoned experiment, nor by the fact that the parts are old, nor by the fact of a prior accidental use or production which is not understood.

Utility is also very well defined. It is precluded if the device be inoperative, or if injurious to the public health, morals or good order.

The word "discovered" as here used is applicable in connection with a process or composition of matter, as for instance, a medical compound, where the value of the novel ingredients as a cure for some particular ill is more in the nature of a discovery.

The word invented is more applicable to a machine, manufacture, etc., and the rules for testing invention will be given further on.

The word art is used in a more narrow sense than ordinarily, Walker's definition being, "an operation performed by rule to produce a result." In this relation it is well to note the same writer's definition of the word process, "an operation performed by rule to produce a result by a means not solely mechanical." A purely mechanical process is not patentable. To be patentable it must be a combined use of several laws of nature, as distinguished from a principle, which is only one law of nature, and hence not patentable.

A process which involves chemical change is patentable. Thus "the manufacturing of fat acids and glycerine from fatty bodies, by the action of water at a high temperature and pressure" would be a patentable process, while "the use of the elec-

6

tric current for marking intelligible signs at any distance" is not patentable.

Another process which has been decided in the courts to involve invention, is that of reducing fibrous sheets to a soft and pliable condition, by first moistening them and then pounding, it being held that the moistening involved chemical change. This decision plainly lies very near the border line of invention, as it would seem that the idea of moistening a substance, when pounding it to get pliability, would be suggested in the mind of any artisan.

A machine is a combination of mechanical parts adapted to receive motion, to produce a desired result. A great majority of patents are upon machines, either designed to effect a new result, or in additions to or subtractions from old machines to facilitate their operation and improve the results obtained thereby.

A composition of matter may be composed of two or more ingredients, mechanically or chemically combined, and must pass the usual tests of invention, novelty and utility. A medical compound, to be patentable, must be more than a mere prescription, in other words some one of its ingredients must have been discovered to be a remedy for an ill to which it has never before been applied.

A manufacture is whatever is made by the hand of man, as distinguished from a machine or composition of matter.

It is not generally necessary before applying for a patent, that it be decided to which of these heads an invention will be assigned.

As to what constitutes the limits of the word "invented" no hard and fast rules can be given. The border line between invention and mechanical skill is very indistinct and its interpretation varies with the ideas of the Examiner or other to whom the question may come to be decided. The lowest order of invention is something more than the highest order of mechanical skill, and the question is not whether a particular mechanic could make the improvement, but whether *any* mechanic might.

Invention may more often be tested by negative rules or the process of elimination. All improvement is not invention. Invention does not lie in superior workmanship, superior materials, in greater strength, in fact it is not determined by degree, although special cases have arisen under each of these headings, in which invention was allowed. For instance, in certain processes, where advantages arose from the treatment of certain substances at a higher temperature than had been before attempted, limited claims were allowed.

A change of material or shape does not constitute invention, unless it be the means of producing a new effect, or accomplishing

7

a new and improved result. Thus the making of solid castings in lieu of the construction of detached parts is not invention.

Aggregation is not invention. As long as each element performs some old and well known function, the result is not a patentable combination but an aggregation of elements. The multiplicity of elements may go on indefinitely without creating a patentable combination, unless by their union a new result be produced. A pertinent example of an aggregation is the addition of the rubber tip to the lead pencil, which was patented, but the patent was declared void by the courts.

To be patentable, the combination of old parts must produce a new result, and the result must be dependent upon the collective union of the old parts. For example, the affixing of the barbs, forming the barbed wire fence, was a patentable combination.

Duplication or multiplication of parts is not invention, nor is the omission of an element, where accompanied by the omission of the function of that element. If, however, there be a new method of operation, whereby the same result is obtained, then the omission of an element may result in a patentable combination.

The substitution of an equivalent is not invention, nor is the use of an old thing for a new purpose unless there is a change in the method of application. As an example of the latter, it has been decided that giving the holes in a gas heater a downward inclination where in the prior art the holes were horizontal, a change which was attended with new and beneficial results, constituted invention.

A new means of producing an old result is generally patentable.

Inventions may be either generic or specific.

There can only be one genus, but there may be several species.

A generic or primary invention is one that performs a function never before performed by any earlier invention. It involves the creation of a new art. The means and also the process may be patented irrespective of any particular means, but one means must be pointed out in the specification.

A specific or secondary invention is one that performs a function previously performed by some earlier invention, but which performs that function in a substantially different way.

The Bell telephone is a primary invention.

It is important to fully understand this distinction, since a primary invention well protected is of the greatest value, and it must be recognized as such, to get suitable protection.

The Patent Office will not answer any inquiries as to the novelty of an alleged invention, nor as to its patentability. As to these points the inventor must judge for himself. The Office and

8

its records are open to him, and if he is not able to make an examination in person, he may either have a search made by experts or he may send for copies of all the patents relating to the art to which his invention belongs. The Office will supply these copies at the rate of 5 cents each, or if ordered by subclasses at the rate of 3 cents each, or by classes at the rate of two cents each. A letter to the Patent Office as to the number and cost of copies of patents in any particular class or sub-class, will obtain the desired information.

3. CORRESPONDENCE.

All business with the Patent Office must be transacted in writing, hence the personal attendance of the applicants is unnecessary, although a personal interview with the examiner, to explain and make more clear a disputed point will sometimes be of advantage, when possible.

All postage and other charges must be prepaid in full.

A registered letter or Post Office money order affords a safe means of transmitting money to the Office.

All letters and communications must be addressed the "Commissioner of Patents."

The assignee of an entire interest in an application is entitled to hold correspondence with the Patent Office to the exclusion of the inventor. When there has been an assigment of an undivided part of the invention all amendments requiring the signature of the inventor must have the assent of the assignee. (See appendix, form 6.)

When the attorney shall have filed his power of attorney, duly executed correspondence will then be held with him.

A double correspondence with the inventor and the attorney will not be allowed.

A separate letter should in every case be written in respect to each distinct subject of inquiry or application.

A letter concerning an application should have the name of the applicant, title of invention, date of filing of application, and serial number of the same at the head of the letter. (See appendix, forms 13 and 18.)

Communications which are wanting in decorum or courtesy will be returned.

4. APPLICATION.

A complete application comprises the first fee of $15.00, Petition, Specification, Oath, and generally drawings; a model or specimen when required by the Office. The application will not be placed on file until all its parts have been received at the Office, when a receipt will be sent to the applicant, giving the filing date and serial number.

It is desirable that all parts should be sent together, otherwise a separate letter must accompany each; and to this end the best mode of transmission is by a pasteboard or other mailing tube, which may be registered to ensure its safe delivery.

The petition, specification and oath should be all attached together in the order named.

When a forfeited application is renewed the original specification, oath and drawing may be used. (See appendix, form 9.)

Papers not amounting to a complete application will be returned upon request.

PETITION.

If a power of attorney is given it is usually made a part of the petition. (See appendix, form 3, also forms 1 to 9 incl.)

SPECIFICATION.

(See appendix, forms 10, 11, 14, 15, 16, 19.)

The specification is a written description of the invention, and of the manner or process of using, constructing or compounding the same, and is required to be in such full, clear and exact terms that any person skilled in the art or science to which it belongs may be enabled to make, use, construct or compound the same.

Persons "skilled in the art" are only those of fair and ordinary information and not those having great technical knowledge of the subject matter of the invention.

The specification must set forth the precise invention for which a patent is desired, explain the principle thereof and the best mode in which the applicant has contemplated applying that principle.

In case the invention is an improvement, the specification must clearly point out the parts to which the improvement relates, and distinguish between what is old and what is new.

11

The specification must conclude with a specific or distinct claim or claims of the part, improvement or combination which the applicant regards as his invention or discovery.

The order to be followed in framing the specification is:

1. Preamble stating the name and residence of the applicant, and title of the invention, a statement as to whether it has been patented in any country, and if so the country in which it has been so patented, and the date and number of each patent.

2. A statement of the general nature and object of the invention.

3. A brief description of the different views of the drawings (where drawings are admissible.)

4. A detailed description of the invention, referring to the drawing by suitable characters, figures being most desirable, since their number is unlimited; the structure is first described, then its mode of operation, and finally what are believed to be the novel and patentable features are briefly distinguished from the old, as in most cases it is necessary to show more or less old construction to make plain the operation.

If the subject is a composition of matter, the description should state the component parts thereof, the proportions in which they are to be mixed, with clearness and precision. (See appendix, form 15.)

A description need not state every use to which an invention may be applied, in order to cover such use, since the applicant is not required to foresee all the advantages of his invention; but the omission of any thing absolutely material to the utility of the invention described, is a fatal defect in the specification, unless that omission would be naturally supplied by anyone skilled in the art.

5. The claims. These form the most vital part of the whole application.

6. Signature (in full) of the inventor.

7. Signature (in full) of two witnesses.

CLAIMS.

Each claim is a more or less concise description of the invention, and interpreted with reference to the drawings and description forms the means of identifying the invention. The more comprehensive the claim the greater the protection it offers, while every limiting word and almost every unnecessary word narrows its scope. Each claim is an entirety and must be interpreted as such. The number of claims to be made depends upon the complexity of the invention and the number of novel points contained therein. Each claim must be designed to protect a particular feature of the invention, and in most cases each feature must be protected by both broad and particular claims. Each claim is made up of elements and the number of elements must be pruned to the barest skeleton which is necessary to a working combination. A claim that would describe a previously existing machine is not allowable, since the inventor would then be claiming something not of his invention, and hence not his property. Thus the scope of the broadest claim in an application is limited by the "state of the art," as it is called, or by the inventions on record relating to this art. If the inventor understand these points he need not fear the bugbear of combination claims, since every claim is a combination claim, and only becomes valueless when a large or unnecessary number of elements is introduced.

Functional claims are: First, those covering a function, result or effect, not a product or composition of matter. These are usually participial in form, as — "imparting a co-existing movement to two reciprocating catch pieces in the operation of trip cut-off valves" — and are not allowable. Second, those covering the function of a machine, as a mechanical process or 'method ; these are not allowable. Third, those reciting the functions of elements and not their structure, as those defining the construction as "means" or "mechanism," etc.; these are allowable except when the novelty of the invention lies in the "means" or "mechanism" so recited. Fourth, those defining sets of mechanism by the results produced. These are allowable if the state of the art permits.

There are two methods of drawing up the claims for an application. One is to use as a foundation claim the broadest claim consistent with the writers knowledge of the art, and taking this as a basis to make other claims, each a little more limited than the previous, until every point is covered, making each claim distinct from every other, either by covering a distinct point, or by covering the same feature more definitely or more broadly, by the addition of or substraction of a detail of construction. As in this case the writer proceeds without an accurate knowledge of the state of the art, the result is that more or less of the claims are rejected on prior inventions, and have to be limited or cancelled. A much better practice, and one requiring no more work in the end in most cases, is to first obtain from the patent office copies of all the patents relating to the subject involved, and after consulting them to frame the claims so as to avoid the prior patents.

The main points to be kept in mind in framing claims thus are : to avoid claiming any prior invention, to make at least one claim so comprehensive that it will cover other possible ways of reaching the end that the inventor has in view, and yet to make the claim definite. It must be kept in mind that no claim may be made which is not illustrated in the drawing or pointed out in the description. Hence, where possible, it is well to show several ways that will effect the desired result, as well as the preferred way; and the inventor ought to direct his thoughts to all other ways of reaching the end sought, and by showing them he may cover possible ways in which another might get around his claims.

In this reference it must be remembered that two distinct inventions can not be claimed in one application. The test as to whether they are distinct or not is to determine whether they may be defined by one claim. If they can then it is allowable to include them in one application, otherwise separate applications must be made. Where several ways of effecting a desired result are thus shown, they are covered by as many broad claims as possible, and then the preferred way is made the subject of specific and detail claims.

Where there are several inventions mutually dependent upon each other and all contribute to a single result, then they may be claimed separately in one application.

If several inventions claimed in one application be of such a nature that they may not be covered by a single claim, the inventor will be required to confine his claims to one, and the rest may be made the subject of separate applications.

Divisional applications may generally be determined by these points : Whether the devices are interdependent, connected in design and operation, capable of use in other relations than that in which they are shown, and have a recognized standing in the art as subjects of separate manufacture and sale. Where there is any doubt in the applicant's mind on this point, he may file one application for all, and then if the Office requires a division, he may do so.

In case an application is required to be divided, a new and complete application must be made for each subject required to be separated, and must include the first fee, petition, specification, oath, and drawing : the relation the new application bears to the old, identifying the latter by its filing date and number, must be stated in the specification.

The specification must in all cases be plainly written, preferably type written, on one side of the paper, with a wide margin of one inch or more on the left hand side. Legal cap with the lines numbered is preferred. An exact copy of all parts of the application, and all amendments should be kept for future reference.

OATH.   (See appendix, forms 20, 21, 22, 23.)

An additional oath will be required if the application is not filed within a reasonable time after the execution of the original oath.

14

The oath must be executed or signed before a Notary Public.

In case an applicant seeks by amendment, to introduce a new claim, not originally described in the specification, he will be required to file a supplemental oath. (See appendix, form 22.) This relates only to the case where a feature not described in the specification is made the subject of a claim. And with the amendment to the claims must be added an amendment to the specification describing the feature defined by the new claim.

DRAWINGS.

The applicant will be required to furnish a drawing whenever the case permits. The drawings must show every feature of the invention which is to be claimed. When the drawing consists of an improvement on an old machine, the drawing must exhibit the invention itself disconnected from the old machine, and also in another view the specific improvement and such parts of the old machine as necessarily co-operate with it.

Pure white paper must be used, 10 x 15 inches in size, of a thickness corresponding with three-ply bristol board, calendared and smooth. The views must be made by the pen only, in India ink, to secure black and solid lines, and the fewest lines consistent with clearness. They must not be too fine or crowded. Surface shading must be open. The different views must be consecutively numbered. The same reference numerals or letters must refer to the same parts throughout all the views. One of the shorter edges must be the top, and a marginal line 1 inch from the edge must be made all around the sheet. A clear space of an inch and a quarter below the top marginal line must be left for the heading which will be put in by the Patent Office in uniform style. Each sheet must be signed by the inventor, or the name of the inventor must be signed upon the drawing by the attorney in fact, at the lower right hand corner, with reference to the heading, attested by two witnesses at the lower left hand corner. All views and signatures must be within the marginal lines, and if the drawing be arranged lengthwise of the sheet, the views and reference numerals must read when the heading is towards the right. The drawings must not be folded for transmission to the office. No new matter can be added to the drawing after filing unless described in the specification.

Blue-print copies of drawings may be obtained from the Patent Office, in three different sixes, the smallest, 5 x 8 inches at five cents a sheet; when ordering this size two copies of each sheet should always be ordered and one copy of each sheet be placed on file with the application. Larger sized copies may be obtained as follows: 7 x 11 inches, fifteen cents a sheet, and 10 x 15 inches twenty-five cents a sheet.

MODEL.

It is very rare that a model is required, so it need be furnished only when demanded by the Patent Office ; this would amount to an official action, and action on the application would be suspended a reasonable time.

The model should exhibit clearly every feature of the invention which it is desired to claim, and must not be more than one foot in length, height or breadth, and must be made of substantial material.

Models made of wood not to be glued up.

SPECIMENS.

When the invention is a composition of matter, the applicant is sometimes required to furnish specimens of the composition, and ingredients sufficient in quantity for the purpose of examination.

5. DESIGN PATENTS.

Revised Statutes of the United States, Section 4929: "A patent may also be obtained by any person who by his own industry, genius, efforts and expense, has invented and produced any new and original design for a manufacture, bust, statue, alto-relievo, or bas-relief; any new and original design for the printing of woolen, silk, cotton and other fabrics; any new and original impression, ornament, pattern, print or picture, to be painted, printed, cast or otherwise placed on or worked into any article of manufacture; or any new, useful and original shape and configuration of any article of manufacture. the same not having been known or used by others before his invention thereof, nor patented nor described in any printed publication, upon payment of the fees required by law.

Evidently the chief requirement for a design patent is novelty.

The novelty of a design is tested, not by any investigation of the means employed for its creation, but by ocular comparison of the design itself with prior designs in the art; and in the eyes of average observers, not experts.

A design is not negatived by the fact that all of its features can be collected out of some scattered designs.

A design patent and a mechanical patent may be granted for the same invention, since the subject matter of the claims can not be the same.

The protection offered by designs patents is very limited, since the claim must be limited to the precise configuration shown.

Design patents are granted for three and a half, seven and fourteen years, and the fees are respectively, $10.00, $15.00, and $30.00.

The proceedings in applications for design patents are substantially the same as in applications for other patents. The specifications must distinctly point out the characteristic features of the design and distinguish between what is old and what is new. The claims should be distinct and specific.

The following order of arrangement is to be observed.

Preamble with name and residence of the applicant, title of the design, and name of the article for which the design has been invented.

Detailed description of the design, as it appears in the drawing.

Claim or claims, Signature of the inventor, and signatures of two witnesses. (See appendix, form 14.)

When the design can be represented by a drawing a model will not be required. The drawing must follow the rules formulated for mechanical drawings. A photograph must be filed when required by the examiner.

It must be mounted upon bristol board of the size and shape required by drawings and similarly signed and witnessed.

6. EXAMINATION.

When all parts of the application have been received at the Patent Office it is given a serial number and a date. The applications as they are received are classified according to the various arts and taken up for examination in the regular order of their filing in their respective classes. It takes a varying length of time for each application to be reached for action but reference to the Official Gazette will show how far behind the Office is in any particular class.

The first step in the examination will be to determine whether the application is in proper form. If, however, the objections are not vital the examiner will proceed with the consideration of the application upon its merits and in such case, in his first letter to the applicant will state his objections, whether formal or otherwise, and until the formal objections are disposed of further action may not be taken upon the merits without order of the Commissioner.

The following applications have precedence over others in the order of examination: 1. Applications which are deemed of peculiar value to some branch of public service. 2. Applications for re-issues. 3. Applications covering inventions patented by the inventor in foreign countries, and applications interfering therewith. 4. Applications which would interfere with another application or with an unexpired patent. 5. Cases remanded by an appelate tribunal for further action by the Office. 6. Amended applications.

REJECTIONS — REFERENCES.

The application having been examined, the examiner in his first letter to the applicant criticises anything which seems to him informal or incomplete in the drawings and specifications, and rejects any claims which may be read of any patent on record, citing the best references on which his rejection is based and fully designating them so as to enable the applicant to look them up, giving their dates, numbers, and the names of the patentees, and the classes of invention.

An applicant will be considered to persist in his claim for a patent without altering his specification in case he fail to act in prosecution of the same for six months after the Office action thereon, and thereupon the examiner will make a second examination of the case.

When a claim in an application is rejected on reference to a domestic patent which shows or describes but does not claim the invention, or on reference to a foreign patent or to a printed publication, and the applicant shall make oath to facts showing a completion of his invention in this country before the filing of the application on which the domestic patent issued, or before the foreign patent was issued, or before the date of the printed publication, then the patent or publication cited shall not prevent the grant of a patent to the applicant.

The examiner may reject a claim for undue breadth, on the ground that while the claim is not met by any prior patent, yet owing to the advanced state of the art, the applicant is not entitled to a claim of much breadth. (See appendix, forms 12, 17.)

AMENDMENTS. (See appendix, forms 13, 18.)

After the applicant has looked up the references on which his claims are rejected, his next step depends on whether he thinks the references are pertinent or not. If they be, he should amend the claims so rejected, either by cancelling them altogether, or by adding limiting words, so as to make the claim avoid the references cited, being careful not to put in any more limiting words than are absolutely necessary. It is also better to accompany the amended claim or claims with a statement or argument in support of them and explaining them, and pointing out their patentable novelty. Sometimes a claim rejected may be written in another form that will more clearly avoid the references without narrowing the scope of the claim.

If the applicant thinks that the claim is just, and not met by the reference, he should ask for a re-consideration of the claim, giving all his arguments in support of it and pointing out the supposed errors of the examiner. If the claim is a second time rejected on the same grounds, the only course open to the applicant is an appeal from the Primary Examiner to the Examiner in chief.

All claims the applicant desires to make must be presented before the appeal stage.

New parties may be admitted to a pending application as joint inventors or some of the parties may be dropped, where they all consent in writing and where the non-joinder or misjoinder was by mistake and without fraudulent intent. And applications may be amended by dividing out distinct though related inventions or by uniting two or more applications where the inventions are related, but not to put the contents of a later application into an earlier one.

In every amendment the matter to be stricken out or inserted must be specified, and the point indicated where the erasure or insertion is to be made.

When an applicant presents a claim for matter originally shown or described but not substantially embraced in the statement of invention or claim originally presented, he will file a supplemental oath to the effect that the subject matter of the proposed amendment was part of his invention, was invented before he filed his original application, was not known or used before his invention, was not in public use or on sale in this country for more than two years prior to the date of his application, and has not been abandoned. (See appendix, form 22.)

During the pendency of an interference, amendments to an application will not be entered, except to withdraw subject matter not adjudged to interfere, or to avoid interference, or to disclaim, or to claim a patentable invention claimed by another party and shown and described by, but not claimed by the party who amends.

After final rejection amendments will not be entered, except to conform to the requirements of the examiner, or to cancel claims, or to present the twice rejected claims and the related parts of the specifications and drawings in better form for consideration on appeal.

After an application has been formally allowed amendments will not be entered except by leave and for a good cause.

If it should become advisable to amend an allowed application and the Office should refuse to entertain the amendment, the application may be forfeited and a new application filed in its place, containing the desired amendment.

The applicant must prosecute his application within two years after the date of the last official action thereon.

ALLOWANCE.

If upon examination it is decided that the applicant is justly entitled to a patent, a notice of allowance will be sent him, calling for the payment of the final fee of $20.00 within six months after the date of the notice, and upon receipt of the final fee within this time, the patent will be issued.

FORFEITED APPLICATION.

In case the final fee is not paid within the prescribed time, the application becomes forfeited, and in order to renew it, a new first fee of $15.00 is required, but the original specification, oath and drawing may be used. (See appendix, form 9.)

ABANDONED APPLICATION.

Any application, including an allowed one, which is not acted upon within two years after the last action by the Office, becomes abandoned, and can only be revived in exceptional cases ; it must be shown to the satisfaction of the Commissioner that such delay was unavoidable.

When an abandoned application is revived a new first fee, petition, specification, oath and drawing will be required.

7. APPEAL.

Every applicant for a patent, any of whose claims have been twice rejected upon the same ground, may upon the payment of a fee of $10.00 appeal from the decision of the primary examiner to the examiner-in-chief. The appeal must set forth in writing the points of the decision upon which it is taken, and must be signed by the applicant or his attorney.

All preliminary and intermediate questions must be settled upon before the case can be appealed.

The appellant must before the day of hearing file a brief of the arguments and authorities upon which he relies to maintain his appeal. If he desires to be heard orally he may so indicate when he files his appeal and a day of hearing will be fixed and due notice given him. In appealable cases in which no limit of appeal is fixed no appeal will be entertained unless within six months after the last action which puts the case in condition for appeal, unless it be shown to the satisfaction of the Commissioner that such delay was unavoidable.

From an adverse decision of the examiner-in-chief an appeal may be taken to the Commissioner in person, upon payment of the fee of $20.00.

8. INTERFERENCES.

An interference is a proceeding instituted for determining the question of priority of invention between two parties claiming substantially the same invention. The fact that one of the parties has already obtained a patent will not prevent an interference, provided the other party can file an affidavit that he made the invention before the patentee's application was filed. Before the interference is declared the inventions of each party to the inter-ference must be decided to be patentable. When an applicant fails to place his application in condition for interference within thirty days, the interference will nevertheless proceed.

The primary examiner will clearly define the issue, designat-ing the claims involved in each application, and will send notices of the interference to each party, naming also the other parties to the interference ; the seniority of the applicants is determined by the filing dates of their respective applications, and each party will be required to file on or before a specified date a preliminary statement of the following facts : (See appendix, form 26.)

1. The date of the original conception of the invention at issue ;

2. The date upon which the first drawing was made ;

3. The date upon which a model of the invention was made ;

4. The date upon which the invention was first disclosed to others ;

5. The date of reduction to practice of the invention ;

6. A statement of the extent of use of the invention.

If a drawing has not been made, or if the invention has not been disclosed to others, or used to any extent, or reduced to practice, the statement must so state.

When the invention was made abroad the statement must set forth :

1. That the applicant made the invention at issue.

2. When and where patented abroad, if so patented.

3. When and where described in a printed publication if so described.

4. When introduced in this country and where.

The statement must be sealed up in a separate envelope be-fore filing, with the name of the party filing it, the subject of the invention and title of the case upon the envelope, which must be enclosed in another envelope for transmission to the Offiee.

﹣ These envelopes will not be opened until statements from all parties to the case have been received or until the date of filing has expired.   And after the approval of the statements submitted, and the date of filing has expired, the parties will be permitted to obtain copies of their opponents cases, at least of as much as pertains to the subject of the interference.   If a party fail to file a preliminary statement his case must be restricted to his date of filing.

When only a part of an application is involved in an interference, if separable, the applicant may file certified copies of that part of the drawings, specifications and claims which relate to the interference and such copies may be used in the proceeding and the remainder of the application withheld from inspection.

There are occasions where it is desirable to avoid an interference by filing a disclaimer, as when the claims involved are of a subordinate nature and the applicant is doubtful of the outcome of the proceedings.

In such case the disclaimer must be filed before the date of filing the preliminary statement, and must disclaim the invention of the particular matter at issue, and cancel the claims involved. (See appendix, form 25.)

It must be kept in mind that an interference is based on conflicting claims when the claims in whole or in part cover the same invention.

Only a small per centage of the whole number of applications for patents meet interference suits, probably not one in twenty-five.   It would not be advisable for the inventor to conduct an interference himself beyond the point of filing his preliminary statement.   The cases are very often decided at this point.

9. RE-ISSUE.

(See appendix, form 21.)

An application for re-issue must be accompanied by the original patent, which will be returned if the re-issue is not granted.

A re-issue will be granted when a patent is inoperative or invalid by reason of an insufficient or defective specification, or by the patentee claiming as his invention more than he had a right to claim, when these errors occurred through no fraudulent or deceptive intent on the part of the patentee.

A claim erased from the original application, through a mistake of the solicitor, may be reinstated in the re-issue.

A claim may be enlarged in a re-issue only in case of a real mistake inadvertently committed.

EXTENSION.

Patents will not be extended except by act of Congress.

10. CAVEATS.

It is generally admitted that the Caveat does not offer the inventor sufficient benefit to ever recompense him for the possible weakening of his case which may follow its filing. It conveys no rights and is always an invitation for an interference suit. It simply is a notice to the Patent Office that the inventor is experimenting with an invention and wants more time to complete it, in other words it puts the inventor on record as having an invention which is not complete at that time, and if some one happens to apply for a patent on the same thing, the caveator finds himself in a weak position, since the act of applying for a patent is prima facie evidence that the invention of the applicant is completed. Moreover, the caveat has no claim upon anything already in the Office, merely being a claim upon the Office to inform the caveator of anything filed after his caveat. Its life is one year and may be renewed. In case of another application being filed for the same thing, the caveator is notified and allowed only three months to complete his invention and make application for a patent, and if he so completes his invention the first result is an interference suit with the application called to his notice, which will have been held three months to await his action.

If the caveator simply wishes to go on record, it is much better to make or get made a sketch or drawing of the application of his idea and have it witnessed by some reputable person and preserve it, and then push his invention to completion as fast as possible.

The fee for a caveat is $10.00. It may be renewed at the end of a year by the payment of another fee.

The caveat comprises an oath, specification and drawing where the case admits of the latter; the drawing may be made on tracing cloth of a size convenient to be folded to the size required for drawings.

The same particularity of description is not required of a caveat as for a patent application, but it must set forth the distinguishing characteristics of the invention.

11. DISCLAIMER.

Disclaimers are of two descriptions; one form is that where a patentee finds his claims cover more than he had a right to claim as new or as his invention; he may in this case make disclaimer of the part not his own on payment of the fee of ten dollars; the other form is that where an *applicant* has filed claims to which he does not choose to claim title; for instance to avoid the continuance of an interference suit, a disclaimer may be filed. Such disclaimers do not require a fee. (See appendix, form 25).

12. ASSIGNMENT.

(See appendix, forms 27 to 32 inclusive.)

An assignment must be recorded in the Patent Office within three months after its date, or it will become void as against a subsequent purchaser. If it is desired to look up the title of a patent or an application, the Office will furnish an abstract of title for a small fee which may be learned by inquiry.

13. APPENDIX OF FORMS.

PETITIONS.

1. BY A SOLE INVENTOR.

To the Commissioner of Patents :

Your petitioner, *James B. Smith*, a citizen of *the United States*, residing at *New York*, in the county of *New York*, and State of *New York*, (or subject, etc.,) prays that letters patent be granted to *him* for the improvement in *sewing machines* set forth in the annexed specification.

James B. Smith.

2. BY JOINT INVENTORS.

To the Commissioner of Patents :

Your petitioners, *James B. Smith* and *John C. Brown*, citizens of *the United States*, residing respectively at *New York*, in the County of *New York*, and State of *New York*, and at *Brooklyn*, in the County of *Kings*, and State of *New York*, (or subjects, etc.,) pray that letters patent may be granted to them, as joint inventors, for the improvement in *washing machines* set forth in the annexed specification.

James B. Smith,
John C. Brown.

3. PETITION WITH POWER OF ATTORNEY.

To the Commissioner of Patents :

Your petitioner, *James B. Smith,* a citizen of *the United States*, residing at *New York*, in the County of *New York*, and State of *New York*, (or subject, etc.) prays that letters patent may be granted to *him* for the improvement in *bicycles* set forth in the annexed specification ; and *he* hereby appoints *Charles H. Wilson*, of the city of *New York*, and State of *New York*, *his* attorney, with full power of substitution and revocation, to prosecute this application, to make alterations and amendments therein, to receive the patent, and to transact all business in the Patent Office connected therewith.

James B. Smith.

4. By an Administrator.

To the Commissioner of Patents :

Your petitioner, *James B. Smith*, a citizen of *the United States*, residing at *New York*, in the County of *New York*, and State of *New York*, (or subject, etc.,) administrator of the estate of *John C. Brown*, late a citizen of *Brooklyn*, deceased (as by reference to the duly certified copy of letters of administration, hereto annexed, will more fully appear,) prays that letters patent may be granted to him for the invention of the said *John C. Brown*, (improvement in *type-writers*) set forth in the annexed specification.

James B. Smith, Administrator, etc.

5. By an Executor.

To the Commissioner of Patents :

Your petitioner, *James B. Smith*, a citizen of *the United States*, residing at *New York*, in the County of *New York*, and State of *New York*, (or subject, etc.,) executor of the last will and testament of *John C. Brown*, late a citizen of *Brooklyn*, deceased (as by reference to the duly certified copy of letters testamentary, hereto annexed, will more fully appear), prays that letters patent may be granted to him for the invention of the said *John C. Brown*, (improvement in *steam engines*) set forth in the annexed specification. *James B. Smith.*

6. For a Re-issue. (By the Inventor.)

To the Commissioner of Patents :

Your petitioner, *James B. Smith*, a citizen of *the United States*, residing at *New York*, in the County of *New York*, and State of *New York*, (or subject, etc.,) prays that *he* may be allowed to surrender the letters patent for an improvement in *pile-drivers*, granted to him *May 16, 1896*, whereof he is now sole owner (or whereof *John C. Brown*, on whose behalf and with whose assent this application is made, is now sole owner, by assignment), and that letters patent may be re-issued to *him* (or to the said *John C. Brown*) for the same invention, upon the annexed amended specification. With this petition is filed an abstract of title, duly certified as required in such cases. *James B. Smith.*

Assent of Assignee to Re-issue.

The undersigned, assignee of the entire (or of an undivided) interest in the above mentioned letters patent, hereby assents to the accompanying application. *John C. Brown.*

7. For Letters Patent for a Design.

To the Commissioner of Patents:

Your petitioner, *James B. Smith*, a citizen of *the United States*, residing at *New York*, in the County of *New York*, and State of *New York*, (or subject, etc.), prays that letters patent may be granted to *him* for the term of *three and one-half* years (or seven years, or fourteen years) for the new and original design for *rugs* set forth in the annexed specification.

James B. Smith.

8. Caveat.

The petition of *James B. Smith*, a citizen of *the United States*, residing at *New York*, in the County of *New York*, and State of *New York*, (or subject, etc., see Caveats), represents:

That *he* has made certain improvements in *cotton-gins*, and that *he* is now engaged in making experiments for the purpose of perfecting the same, preparatory to applying for letters patent therefor. *He* therefore prays that the subjoined description of *his* invention may be filed as a caveat in the confidential archives of the Patent Office.

James B. Smith.

9. For the Renewal of a Forfeited Application.

To the Commissioner of Patents:

Your petitioner, *James B. Smith*, a citizen of *the United States*, residing at *New York*, in the County of *New York*, and State of *New York*, (or subject, etc.), represents that on *June 9, 1896, he* filed an application for letters patent for an improvement in *printing presses*, serial number *534,453*, which application was allowed *August 15, 1896*, but that he failed to make payment of the final fee within the time allowed by law.

He now makes renewed application for letters patent for said invention and prays that the original specification, oath, drawings and model may be used as a part of this application.

James B. Smith.

SPECIFICATIONS.

10. For an Art or Process.

To all whom it may concern:

Be it known that I, *James B. Smith*, a citizen of *the United States*, residing at *New York*, in the County of *New York*, and

State of *New York* (or subject, etc.), have invented certain new and useful improvements *in purifying and increasing the illuminating power of gas without appreciable loss of bulk* (for which I have received letters in England, No. 1,897, dated August 7, 1896), and I do hereby declare that the following is a full, clear and exact description of the invention, which will enable others skilled in the art to which it appertains to use the same.

Heretofore gas has been purified by passing it through animal charcoal ; but when this is used alone, after a short time it loses its power of absorbing impurities, and must then be washed with steam or water, or have atmospheric air blown through it, or be revivified by heat. Used alone, animal charcoal also reduces the candle-power of the gas passed through it and diminishes its bulk.

The object of my invention is to thoroughly purify illuminating gas, to make the operation continuous, and to purify the gas without detracting from its illuminating power, and without causing any appreciable diminution in bulk ; and to this end my invention consists in increasing the power of animal charcoal to eliminate from illuminating gas those substances which are considered impurities, in charging the charcoal with a substance which will prevent it from depriving the gas of illuminants, and in passing the gas to be purified with atmospheric air through the animal charcoal.

To carry my invention into effect, I moisten the charcoal (which may be either new or spent) with coal-tar, or with coal-tar and water, or in some cases with water only, and then charge this mass into one or more vessels, which then constitutes the purifier. I may put the mass into the vessels while still wet, or, unless water alone is used, after it has dried. Through these vessels the gas is to pass ; but before it is admitted I introduce into it at the retorts, or at the stand-pipe or mains beyond, in order to insure a thorough mixture, a small quantity of atmospheric air — say from eight-tenths to two and a half per cent. of the bulk of the gas to be purified. The quantity of air will depend directly upon the impurities of the gas. Any suitable mixing device for thoroughly mixing the admitted air with the gas may be located at any point in the mains between the retorts and the bone-black purifiers, or even at the point of admission to the purifier. For water-gas the charcoal wet with water alone will suffice, air being introduced into the gas.

The oxygen of the air partly unites with the sulphur to form soluble salts, and the rest combines totally with the hydrogen of the sulphureted and other hydrogen sulphur compounds to form water, and part of the sulphur of the sulphureted hydrogen and other sulphur compounds is precipitated in a free state in the charcoal, while the nitrogen partly goes to form, with part of the remaining hydrogen, ammonia bases. No free oxygen passes off with the purified gas, while if any nitrogen goes over, the quantity is so small that it is not detrimental.

By the application of air in this manner the process is rendered continuous, as the charcoal is kept constantly active for a great length of time.

When the absorbing power of the charcoal finally becomes exhausted, it may either be sold for the valuable ammoniacal salts it contains, or it may be revivified, or be washed and freed from sulphur by a suitable sulphur solvent for re-use.

By charging the bone-black with coal-tar I prevent it from taking out of the gas any olefiant gas or other heavy hydrocarbons serving as illuminants.

I may treat the black, either before or after putting it into vessels, as may be most convenient or suitable, with any substance correlative to the illuminants of the gas ; that is, with any substance which will impregnate the black in such a manner that it will not take up such illuminants. I have particularly described coal-tar because that is most readily at hand ; but its hydrocarbon distillates or the benzole series will answer.

In the case of coal-gas, not only is the sulphureted hydrogen with which it is contaminated taken up, as just described, but the illuminating power of the gas, which is somewhat reduced if passed through dry animal charcoal or bone black, is not decreased when air is used and the animal charcoal or bone-black is wet with tar, but is actually improved, inasmuch as it gives a whiter flame, of the same candle-power as the gas not passed through aninal charcoal or bone-black at all.

By the old method, when purification was effected by the use of lime, the sulphureted hydrogen and carbonic acid were absorbed by the lime, and the result was, of course, a loss of the original bulk of the gas. Now, by my process the sulphur and hydrogen are separated, the sulphur remaining in the charcoal and the hydrogen passing through with the gas, while the carbonic acid passes through entire ; and although it passes through unchanged, it is sufficiently carbureted not to detract from the illuminating power of the gas. I thus have practically the same bulk of gas after purification as before this operation, and loss is prevented without detriment to the consumer.

A striking advantage of my process is, that it unites the scrubbing and purifying operations, for the gas may be passed directly from the condenser into my purifiers.

To eliminate sulphureted hydrogen, I may also mix with the charcoal a substance which will of itself decompose sulphureted hydrogen contained in gas—such as oxide of iron, tin, manganese ore, etc.

33

When the gas issues from my purifiers it is entirely free from ammoniacal and sulphur compounds, and is nearly inodorous.

When the charcoal is removed from the purifiers it is also inodorous, and is in no sense offensive and disgusting like gas-lime.

Having fully described my invention, what I claim, and desire to secure by letters patent, is —

1.   In the purification of illuminating gas by means of animal charcoal, the process of preventing absorption of illuminants of the gas by the charcoal, which consists in supplying the charcoal with a suitable correlative to such illuminants, as described.

2.   The process of purifying illuminating gas, which consists in mixing the same with air and then passing it through animal charcoal impregnated with coal-tar, all substantially as described.

James B. Smith.

Witnesses:

Charles Johnson,
William G. Anthony.

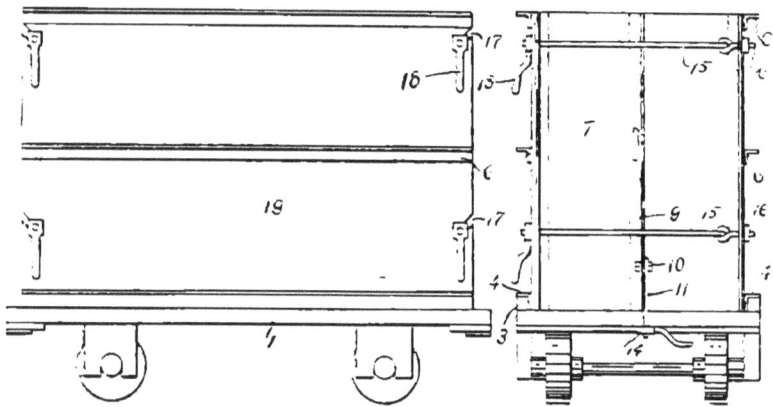

FIG 1.

FIG 2.

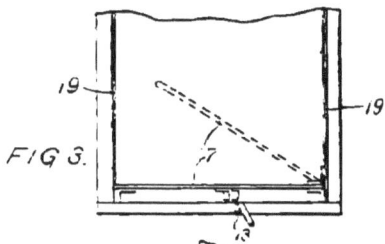

FIG 3.

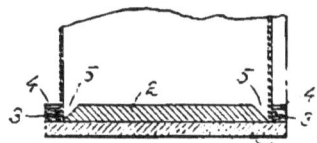

FIG 4.

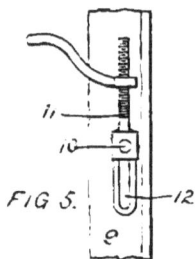

FIG 5.

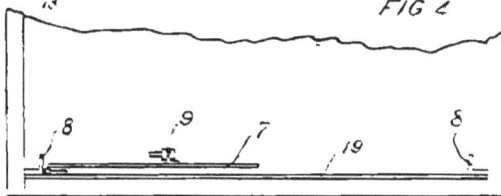

FIG. 6.

WITNESSES
O. L. Plumtre
Anthony Johnson

INVENTOR
James B. Smith
By John C. Brown, Atty.

11. Article of Manufacture.

To all whom it may concern:

Be it known that I, *James B. Smith,* a citizen of *the United States,* residing at *New York,* in the County of *New York,* and State of *New York,* (or subject, etc.,) have invented certain new and useful improvements in *soap frames* of which the following is a full and complete specification.

The object of my improvement is to provide a frame or vat, into which soap may be poured while in a liquid state, to hold it during the process of cooling and hardening, and a frame which can be readily taken apart to remove the soap.

Frames to be adapted to this purpose have to have sufficient capacity to hold about 1000 or 1500 pounds of soap, and when this cools it makes a very unwieldy block, hence it is necessary that the sides and ends of the frame should be capable of being taken down to remove the block.

The frames heretofore used for this purpose have been chiefly of wood and divided into five separate parts, two sides, two ends, and the bottom. To set up and take down a frame of this description is hence a task of some difficulty, especially as the soap always sticks to the sides and makes it difficult to remove them without crumbling the block.

In the drawings Figure 1 is a side elevation of my device. Figure 2 is an end elevation of the same. Figure 3 is a partial plan showing one end. Figure 4 is a section of the base. Figure 5 is a detail of the device for clamping the ends to the base. Figure 6 is a plan of one side with the end folded up.

Referring to the drawings, my improvement has a base portion which consists of a main block 1. Above this is fastened another block 2 of smaller dimensions than the first. The edges of this upper block are bevelled off as shown in Fig. 4, the object of which will be referred to later on. Fastened to the perimeter of the block 1, are the strips 3, fitted to the angle irons 4; the inner vertical face of the angle iron forms with the bevelled edge of the block 2, a groove 5, extending around the base parallel to its sides or ends as the case may be. For each side of the soap frame I provide a plate 19, strenthened by a number of angle irons 6. One end plate 7 is hinged to one end of each side plate. Riveted on the inner side, to each end of each side plate 19, are vertical angle irons 8, against which the end plate 7 rests when the frame is set up. As a means of clamping the end plate to

37

the base, I provide an angle iron 9 which is vertically riveted to the end plate, one flange of this angle projecting perpendicularly to the end plate. A pin or bolt 10 is fastened to this flange and on this bolt is fitted another bolt 11, having a long eye 12, through which the bolt 10 passes. The base is provided with a slot 13 in which the bolt 11 is placed and the nut 14 serves to clamp the end firmly to the base. The object of the eye 12 will be referred to further on. To clamp the sides together, I provide the eye bolts 15, fastened to one side, and having other eye bolts 16 adapted to be fitted into a slot 17 on the opposite side, and having a hand nut 18 to clamp the whole structure together.

In the operation of the frame, after the sides and ends have been placed in their respective grooves, and the end and base bolts tightened up, the soap in a liquid state is poured into the interior. Its weight serves to push the end plates against the angles on the ends of the sides, and at the base to push the ends and side against the flat part of their corresponding grooves, this making in each case a tight joint. After the soap has cooled sufficiently and become hard, the nuts are loosened and the end bolts released, the ends then are sprung outwardly enough to free them from the block of soap, and then the sides are tipped outward from the top. The object of the shape of the grooves in which the sides rest will now be apparent, since the bevel of the inner face of the groove allows the bottom of the side to swing in slightly as its top is pushed outward, thus avoiding the necessity of lifting the sides out of the groove and crumbling the soap. After the sides have been removed and the ends folded up, the bolt 11 is turned on the bolt 10 and allowed to drop down the full length of the slot 12 until it is in the position as shown in Fig. 5, when it lays up against the end plate and can not fall outward until it is again lifted up to the end of the slot.

Having described my invention I claim :

1. In a soap frame, the combination of a base portion, two side plates each provided with an angle iron at each end, and an end plate hinged to one end of each side plate and adapted to swing in towards said side plate, and end angle irons adapted to support said end plates, as and for the purpose set forth.

2. In a soap frame the combination of a base portion provided with a groove parallel to and just within its perimeter, two side plates each provided with an angle iron at each end, an end plate hinged to one end of each side plate and adapted to be supported by one of said end angle irons, said end and side plate adapted to be received in said groove, eyebolts adapted to bind said side plates together, and bolts adapted to bind said end plates to said base, as and for the purpose set forth.

3. In a soap frame, the combination of a base portion, two side plates each provided with an angle iron at each end, an end plate hinged to one end of each side plate and provided with a projecting flange, a bolt fastened to said flange by means of a pin working in a slot in one end of said bolt, as and for the purpose set forth.

James B. Smith.

Witnesses :
Charles Johnson,
William G. Anthony.

12.   Action by the Patent Office.

Department of the Interior,
United States Patent Office,
Washington, D. C., May 10, 1896.
James B. Smith, care of John C. Brown,
New York, N. Y.

Please find below a communication from the Examiner in charge of your application, No. 584,324. Soap Frames. Filed Mar. 10, 1896.

John S. Seymour,
Commissioner of Patents.

Above case has been considered.

Claim 1 is rejected on patent to Babbitt, 342,320, Sep. 21, 1886 (Soap Frames).

Claim 2 is rejected on Rosenblatt, 349,558, May 25, 1886 (same class).

The changes called for by applicants' claims over the above patents are thought to amount to mere mechanical skill.

13.   Amendment.

James B. Smith,
Soap Frame,
Filed Mar. 10, 1896.
Serial No. 584,324.

New York, N. Y., May 15, 1896.
Commissioner of Patents,
Washington, D. C.

Dear Sir :—In the matter of the above mentioned application, and with reference to Office letter of the 10th inst., in order to more particularly point out the patentable distinction of this application from the cases cited against it, viz.: Babbitt, 342,320, Sep.

21, 1886, and Rosenblatt, 349,558, May 25, 1886, I desire to amend as follows:

By striking out claim 1 and substituting the following claim 1.

1. In a soap frame, the combination of a base portion, two side plates each provided with a vertical flange at each end, end plates flexibly joined to said side plates, said flexible joints being within said end flanged whereby said end flanges are adapted to support said end plates when the frame is set up in its normal position, as and for the purpose set forth.

I would respectfully ask for a reconsideration of claim 2 in view of the following facts:

The main and distinctive feature of this invention is the hinged or flexible connection of the sides to the ends. I am aware that the simple addition of such a connection to the structures cited against this application would not ordinarily constitute invention. But in this case a new result has been aimed at and achieved, which it would seem is worthy of protection. In the operation of these frames the cost for the labor of setting them up to receive the liquid soap, and of taking them down after the soap has cooled, in order to remove the hardened block, is a matter which constitutes no small per centage of the cost of production. In the common method of production the frame consists of five parts, a base, two sides and two ends. The difficulty one person would have in setting one of these frames up alone will be readily apparent. In order to avoid this difficulty the applicant invented a frame in which there are only three parts, a base and two sides, the ends of which are flanged to form the ends of the frame, and has already applied for a patent on this construction. At first this construction seemed to meet all requisites, but it was afterwards found that where large quantities of these frames were in use they would take up too much room when not set up. It was therefore to meet this objection that the applicant devised this construction, which when not in use takes up no more room than where the sides and ends are entirely disconnected, and yet which can be set up in half the time. And it is not merely the addition of the flexible joint which is the subject of this application but with that joint a means must be provided of making it tight. This the applicant has ingenuously effected by providing a flange outside of the joint so that the pressure of the soap while in a liquid state will spring the end plates against these flanges and thus make the joint.

It is difficult to fully set forth the advantages of this invention on paper. One ought to have worked at setting up a few frames of the old style and then try one of the applicant's invention to fully appreciate the importance it has.

Respectfully submitted,

JOHN C. BROWN,

Attorney for Applicant.

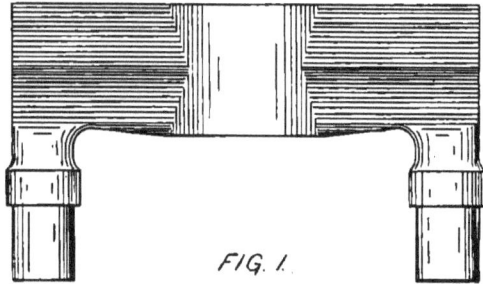

FIG. 1.

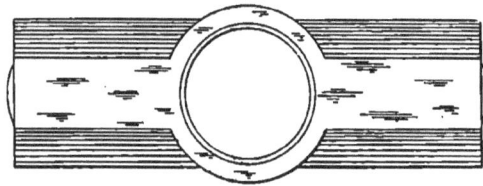

FIG 2

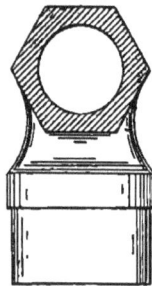

FIG 3.

14. For a Design.

To all whom it may concern :

Be it known that I, *James B. Smith*, a citizen of *the United States*, residing at *New York*, in the County of *New York*, and State of *New York*, (or subject, etc.,) have invented and produced a new and original design for *fork crowns for bicycles*, of which the following is a complete and full specification, reference being had to the accompanying drawings in which ·

Figure 1 is a front elevation of a fork crown embodying my design. Figure 2 is a sectional elevation of the same, and Figure 3 is a plan view of the same.

Referring to the drawings, the leading feature of my design consists in the novel shape which I give to the crown, it being in cross section an approximate hexagon as shown in Figure 2, the angles of the hexagon gradually disappearing near the middle portion of the crown where it is enlarged.

Having described my invention I claim :

The design for a fork crown for bicycles herein shown and described.

James B. Smith.

Witnesses :
Charles Johnson,
William G. Anthony.

15. For a Composition of Matter.

To all whom it may concern :

Be it known that I, *James B. Smith*, a citizen of *the United States*, residing at *New York*, in the County of *New York*, and State of *New York*, (or subject, etc.,) have invented a new and useful composition of matter *to be used for the removal of hair and grease from hides preparatory to tanning*, of which the following is a full and complete specification.

My composition consists of the following ingredients, combined in the proportions stated, viz.:

Pure water 500 gallons, unslaked lime 32 gallons, soda ash 100 pounds, saltpetre 20 pounds, and flowers of sulphur 10 pounds.

These ingredients are to be thoroughly mingled by agitation.

In using the above named composition the hides should first be freed from all salt and impurities, by soaking green hides one day and dry hides eight days, and placing the hides so cleaned in the said solution and allowing them to remain in it for forty-eight hours. The hides are then to be removed from the solution and unhaired in the usual way.

By the use of this composition the hair is speedily and thoroughly loosened, and the hides, while retaining all that portion of the substance which can be converted into leather, are at the same time thoroughly cleaned from grease and other substances which would prevent them from being tanned quickly.

I claim :

The herein described composition of matter to be used for depilating hides and preparing them for being tanned, consisting of water, unslaked lime, soda ash, saltpetre, and flowers of sulphur in the proportions specified.

James B. Smith.

Witnesses :
Charles Johnson,
William G. Anthony.

16. For a Medical Compound.

To all whom it may concern :

Be it known that I, *James C. Smith*, a citizen of the *United States*, residing at *New York*, in the county of *New York* and State of *New York* (or subject, etc.), have invented and discovered a new and useful medical compound, of which the following is a full and complete specification.

My discovery has for its object to provide a new and efficient cure for the dropsy and kindred kidney troubles, and consists of the following ingredients : Apocynum (Indian Hemp), Gaultheria Procumbens (Wintergreen), and Polytrichum Communis, (Golden Moss) ; the latter is a moss which is found upon rocks in shaded places in the mountains of Virginia, and forms the novel and most efficient ingredient of the compound.

In preparing the compound I first take equal parts of Indian Hemp and Golden Moss, about two ounces of each, and steep in about three quarts of water for six hours, reducing the whole volume to about two quarts or to about two-thirds of the original volume. I then add one-half dram of wintergreen to each quart of the compound. Wintergreen being an astringent, is used to counteract the too severe laxative effect which the two other ingredients would otherwise have upon the bowels and kidneys.

The proper dose is a wine glass full taken two or three times a day. The compound when made in the manner hereinbefore described forms a positive and never failing cure for the dropsy and kindred troubles.

Having described my invention I claim :

A medical compound composed of an aqueous solution of

44

Apocynum, Gaultheria Procumbens and Polytrichum Communis, in the proportions herein set forth. *James B. Smith.*

Witnesses :
Charles Johnson,
William G. Anthony.

17. Action by the Patent Office.

United States Patent Office.

Washington, D. C., Oct. 10, 1895.

James C. Smith
New York, N. Y.

Please find below a communication from the Examiner in charge of your application, Medical Compound, filed August 16, 1895, Serial No. 565,789.

John S. Seymour,
Commissioner of Patents.

It is admitted by the examiner that a compound consisting of the Apocynum and Wintergreen would form a useful compound for the treatment of dropsy and allied diseases. Whether the addition thereto of Polytrichum Communis would add to or detract from the curable properties of the first named diuretic is question-able. The use of the first two ingredients of this compound does not exceed the expected skill of the medical profession, and is not a discovery within the meaning of the law. It amounts simply to a prescription and prescriptions are not patentable. It is obvious that there are many herbs not on record as useful for the treat-ment of any disease, which if added to the compound first named, would not interfere with its diuretic action, but still, while such addition would give numerical novelty, it would not make up a patentable combination.

Further action on this case is therefore suspended to afford the applicant time to demonstrate the value of his alleged new remedy, by experiment, apart from and not in connection with the first two named ingredients.

18. For an Amendment.

James B. Smith,
Medical Compound,
Filed Aug. 16, 1895.
Serial No. 565. 789.

New York, N. Y., Oct. 15, 1895.

Commissioner of Patents,
Washington, D. C.   .

Dear Sir :—In the matter of the above mentioned application and with reference to Office letter of the 10th inst., I desire to amend as follows :

By inserting between paragraphs 2 and 3, page 1 of the specifications, the following paragraph :

I am aware that the Apocynum has been proposed before for use in the treatment of the above mentioned diseases, but in experiments made with the Polytrichum Communis apart from the Apocynum the value of each has been verified, and it has been found that the absence of either of these ingredients detracts from the curable properties of the compound.

I would respectfully ask for favorable action on this case in view of the following facts : That this moss has been shipped to me in large quantities, at considerable expense to have it preserved in proper condition. I would not go to this trouble and expense if any other ingredient would take the place of the Polytrichum Communis. It is admitted that the Apocynum is useful in dropsy cases as it is so given in various medical works. But the fact that the compound as described in the original specification will cure aggravated cases, which the Apocynum will not cure, and the fact that the Polytrichum Communis is not on record as useful in such cases would seem to entitle me to protection on the compound as described in the specification.

Respectfully Yours,
James B. Smith.

19. FOR A CAVEAT.

To the Commissioner of Patents :

Be it known that I, *James B. Smith,* a citizen of *the United States,* residing at *New York,* in the County of *New York,* and State of *New York,* (or subject, etc.,) having invented an improvement in *velocipedes,* and desiring further to mature the same, file this my caveat therefor, and pray protection of my right until I shall have matured my invention.

The following is a description of my newly invented *velocipede,* which is as full, clear and exact as I am at this time able to give, reference being had to the drawing hereunto annexed.

The object of my invention is to render it unnecessary to turn the front wheel so much as heretofore, and at the same time to facilitate the turning of sharp curves. This I accomplish by fitting the front and rear wheels on vertical pivots, and connecting them by means of a diagonal bar, as shown in the drawing, so that the turning of the front wheel also turns the back wheel, thereby enabling it easily to turn a curve.

In the drawing 1 is the front wheel, 2 the rear wheel, and 3 the standards extending from the axle of the front wheel to the vertical pivot 5 in the beam 6, and 4 is the cross-bar upon the end

5, by which the steering is done. The rear wheel 2 is also fitted with jaws 7 and a vertical pivot 8.

James B. Smith.

Witnesses:
Charles Johnson,
William G. Anthony.

OATHS.

20. BY AN INVENTOR.

State of *New York,* } ss.
County of *New York,* }

James B. Smith, the above named petitioner, a citizen of *the United States,*[1] and resident of *New York,* in the County of *New York,* and State of *New York,* being duly sworn, deposes and says that *he* verily believes *himself* to be the original, first and *sole*[2] inventor of the improvement in *bicycles* described and claimed in the foregoing specification; that the same has not been patented to *him,* or to others with *his* knowledge or consent, *in any country;*[3] that the same has not to *his* knowledge been in public use or on sale in the United States for more than two years prior to *his* application, and *he* does not know and does not believe that the same was ever known or used prior to *his* invention thereof.

James B. Smith.

Subscribed and sworn to before me this *fifteenth* day of *March, 1897.*

John C. Brown,
[Seal]   Notary Public.

21. BY AN APPLICANT FOR A RE-ISSUE (INVENTOR).

State of *New York,* } ss.
County of *New York,* }

James C. Smith, the above named petitioner, being duly sworn, deposes and says that *he* does verily believe *himself* to be the original and first inventor of the improvement set forth and claimed in the foregoing specification and for which improvement *he* solicits a patent; that deponent does not know and does not believe that said improvement was ever before known or used; that deponent

---

[1] If the applicant be an alien, he will state of what foreign or sovereign State he is a citizen or subject.

[2] "Sole" or "joint."

[3] Or "except in the following countries," giving the name of the countries in which it has been so patented, and the date and number of each patent.

is a citizen of *the United States of America*, and resides at *New York*, in the County of *New York*, and State of *New York;* that deponent verily believes that letters patent referred to in the foregoing petition and specification and herewith surrendered are inoperative (or invalid), for the reason that the specification thereof is defective (or insufficient), and that such defect (or insufficiency) consists particularly in   ; and deponent further says that the errors which render such patent so inoperative (or invalid) arose from inadvertence (or accident, or mistake), and without any fraudulent or deceptive intention on the part of deponent; that the following is a true specification of the errors which it is claimed constitute such inadvertence (or accident, or mistake), relied upon ;   ; that such errors so particularly specified arose (or occurred) as follows :

James C. Smith.

Subscribed and sworn to before me this *tenth* day of *May*, *1896*.

John B. James,
Notary Public.

[Seal]

22. Supplemental Oath to Accompany a New and Enlarged Claim.

State of *New York*,
County of *New York*, } ss.

James B. Smith, whose application for letters patent for an improvement in *harvesting machinery*, (serial number 543,345), was filed in the United States Patent Office on or about the *16th* day of *March, 1897*, being duly sworn, deposes and says that *he* verily believes *himself* to be the original and first inventor of the improvement as described and claimed in the foregoing amendment, in addition to that which was embraced in the claims originally made, and that *he* does not know and does not believe that the same was ever before known or used, and that the matter sought to be inserted formed a part of *his* original invention at the date of filing said application, and was invented by *him* before *he* filed the same.

James B. Smith.

Sworn to and described before me this *tenth* day of *April*, *1897*.

John C. Brown,
Notary Public.

[Seal]

23. Oath as to the Loss of Letters Patent.

State of *New York*,
County of *New York*, } ss.

James B. Smith, of said county, being duly sworn, doth depose and say that the letters patent No. 234,554, granted to him, and bearing date on the *ninth* day of *January, 1897*, have been either lost or destroyed; that he has made diligent search for the said letters patent in all places where the same would probably be found, if existing, and that he has not been able to find them.

James B. Smith.

Subscribed and sworn to before me this *fifth* day of *June, 1896.*

John C. Brown,
[Seal] Notary Public.

24. Revocation of Power of Attorney.

To the Commissioner of Patents:

The undersigned having, on or about the *24th* day of *November*, 1896, appointed *John C. Brown*, of *New York*, in the county of *New York*, and State of *New York*, his attorney to prosecute an application for letters patent, which application was filed on or about the *28th* day of *November*, 1896, for an improvement in *fire engines*, hereby revokes the power of attorney then given.

Signed at *New York*, in the county of *New York*, and State of *New York*, this *23d* day of *December*, 1896.

James B. Smith.

25. Disclaimer during Interference.
Interference. No. 18,879.

James B. Smith,
John C. Brown, } Before the Examiner of Interferences.

Subject matter: *Bicycles.*

To the Commissioner of Patents:

Sir:—In the matter of the interference above noted, under the provisions of and for the purpose set forth in Rule 107. I disclaim: "In an adjustable handle bar, the combination of a bar having a worm gear suitably connected with it, a socket fitted to said bar and adapted to be fitted to the front head of a bicycle, and a worm engaging with said worm gear, as and for the purpose set forth." As I am not the first inventor thereof, I herewith transmit an amendment to my application (serial number

601,768), for the purpose of having the above disclaimer embodied as a part of my specification.

Signed at *New York*, in the county of *New York*, and State of *New York*, this 16th day of *June*, 1896.

John C. Brown.

Witnesses :
Charles Johnson,
William G. Anthony.

26. PRELIMINARY STATEMENT FOR DOMESTIC INVENTOR.

James B. Smith, } Interference in the United States Patent
Office,
John B. Williams, } Preliminary Statement of *James B. Smith.*

James B. Smith, of *New York*, in the County of *New York*, and State of *New York*, being duly sworn, doth depose and say that he is a party to the interference declared by the Commissioner of Patents, *June* 4, 1896, between *James B. Smith's* application for letters patent, filed *April* 3, 1896, serial number 543,345, and the patent to *John B. Williams*, granted *March* 1, 1896, for a *twine machine ;* that he conceived the invention contained in claims 1 and 3 of his application declared to be involved in this interference, on or about the *3d* day of *June*, 1894 ; that on or about the *5th* day of *June*, 1894, he made drawings of the invention (if he has not made a drawing he should so state) ; that on or about the *7th* day of *June*, 1894, he first explained the invention to others, and that he made a model showing such invention on or about the *20th* day of *June*, 1894 (if he has not made a model he should so state) ; that he embodied his invention in a full sized machine, which was completed about the *5th* day of *July*, 1894, and that on the *6th* day of *July*, 1894, said machine was successfully operated at his shop in the city of New York, County of New York, and State of New York, and that he has since continued to use the same, and that he has manufactured others for use and sale. (If he has not embodied the invention in a full sized machine he should so state. And if he has so embodied it but not used it he should so state.)

James B. Smith.

Subscribed and sworn to before me this *12th* day of *June*, 1896.

Charles O. Johnson,
Notary Public.

ASSIGNMENTS.

27. OF AN ENTIRE INTEREST IN AN INVENTION BEFORE THE ISSUE OF LETTERS PATENT.

Whereas I, *James B. Smith*, of *New York*, County of *New York*, State of *New York*, have invented a certain new and useful improvement in *harvesters* (giving title of the same), for which I am about to make application for letters patent of the United States; and whereas *John C. Brown*, of *Brooklyn*, County of *Kings*, State of *New York*, is desirous of acquiring an interest in said invention, and in the letters patent to be obtained therefor:

Now, therefore, to all whom it may concern, be it known that for and in consideration of the sum of *one* dollar to me in hand paid, the receipt of which is hereby acknowledged, I, the said *James B. Smith*, have sold, assigned and transferred, and by these presents do sell, assign and transfer unto the said *John C. Brown*, the full and exclusive right to the said invention, as fully set forth and described in the specification prepared and executed by me on the 15*th* day of *June*, 1896, preparatory to obtaining letters patent of the United States therefor; and I do hereby authorize and request the Commissioner of Patents to issue the said letters patent to the said John C. Brown as the assignee of my entire right, title and interest in and to the same, for the sole use and behoof of the said John C. Brown and his legal representatives.

In testimony whereof I have hereunto set my hand and affixed my seal this 4*th* day of *July*, A. D. 1896.

James B. Smith.   [Seal]

In presence of
Charles Johnson,
William G. Anthony.

28. OF THE ENTIRE INTEREST IN LETTERS PATENT.

Whereas I, *James B. Smith*, of *New York*, County of *New York*, State of *New York*, did obtain letters patent of the United States for an improvement in *car wheels*, which letters patent are numbered 543,345, and bear date the 5th day of *June*, in the year 1896; and whereas I am now the sole owner of said patent and of all rights under the same; and whereas *John C. Brown*, of *Brooklyn*, County of *Kings*, State of *New York*, is desirous of acquiring the entire interest in the same.

Now, therefore, to all whom it may concern, be it known that, for and in consideration of the sum of one dollar to me in hand paid, the receipt of which is hereby acknowledged, I, the said *James B. Smith*, have sold, assigned, and transferred, and by these presents do sell, assign, and transfer unto the said *John C. Brown* the whole, right title and interest in and to the said improvement in *car wheels* and in and to the letters patent therefor aforesaid : the same to be held and enjoyed by the said *John C. Brown*, for his own use and behoof, and for the use and behoof of his legal representatives, to the full end of the term for which said letters patent are or may be granted (thus including extension), as fully and entirely as the same would have been held and enjoyed by me had this assignment and sale not been made.

IN TESTIMONY WHEREOF I have hereunto set my hand and affixed my seal at *New York*, in the County of *New York*, and State of *New York*, this 25th day of *July*, A. D. 1896.

James B. Smith. [Seal]

In presence of
Charles Johnson.
William G. Anthony.

29. OF AN UNDIVIDED INTEREST IN LETTERS PATENT.

Whereas I. *James B. Smith*, of *Brooklyn*, County of *Kings*, State of *New York*, did obtain letters patent of the United States for an improvement in *hay rakes*, which letters patent are numbered 543,216, and bear date the 3d day of *August*, in the year 1895, and whereas *John C. Brown*, of *Brooklyn*, County of *Kings*, State of *New York*, is desirous of acquiring an interest in the same :

Now, therefore, to all whom it may concern, be it known that, for and in consideration of the sum of *one* dollar to me in hand paid, the receipt of which is hereby acknowledged, I, the said *James B. Smith*, have sold, assigned, and transferred, and by these presents do sell, assign and transfer unto the said *John C. Brown* the undivided one-half part of the whole right, title and interest in and to the said invention and in and to the letters patent therefor aforesaid : the said undivided one-half part to be held and enjoyed by the said *John C. Brown*, for his own use and behoof, and for the use and behoof of his legal representatives, to the full end of the term for which said letters patent are or may be granted (thus including extension), as fully and entirely as the same would have been held and enjoyed by me had this assignment and sale not been made.

In testimony whereof I have hereunto set my hand and affixed my seal at *Brooklyn*, in the County of *Kings*, and State of *New York*, this 7th day of *June*, A. D. 1896.

James B. Smith. [Seal]

In presence of
Charles Johnson.
William G. Anthony.

30. Territorial Interest After Grant of Patent.

Whereas I, *James B. Smith*, of *Brooklyn*, County of *Kings*, State of *New York*, did obtain letters patent of the United States for an improvement in *grain-binders*, which letters patent are numbered 345,453, and bear date the *8th* day of *June*, in the year 1895; and whereas I am now the sole owner of said patent and of all rights under the same in the below recited territory; and whereas, *John C. Brown* of *Brooklyn*, County of *Kings*, State of *New York*, is desirous of acquiring an interest in the same:

Now, therefore, to all whom it may concern, be it known that for and in consideration of the sum of *one* dollar to me in hand paid, the receipt of which is hereby acknowledged, I, the said *James B. Smith*, have sold, assigned and transferred, and by these presents do sell, assign and transfer unto the said *John C. Brown*, all the right, title and interest in and to the said invention, as secured to me by said letters patent, for, to, and in the *State of New Jersey*, and for, to, or in no other place or places; the same to be held or enjoyed by the said *John C. Brown*, within and throughout the above specified territory, but not elsewhere, for his own use and behoof, and for the use and behoof of his legal representatives, for the full end of the term for which said letters patent are or may be granted, as fully and entirely as the same would have been held and enjoyed by me had this assignment and sale not been made.

In testimony whereof I have hereunto set my hand and affixed my seal at *Brooklyn*, in the County of *Kings*, and State of *New York*, this 3*d* day of *July*, 1896.

Witnesses: *James B. Smith.*   [Seal]
Charles Johnson,
William G. Anthony.

31. License, Shop-Right.

In consideration of the sum of *one* dollar, to be paid by the firm of *Smith, Brown & Co.*, of *Brooklyn*, in the County of *Kings*, State of *New York*, I do hereby license and empower the said *Smith, Brown & Co.* to manufacture in said *Brooklyn* (or other place agreed upon) the improvement in *cotton-seed planters*, for which letters patent of the United States No. 425,321, were granted to me *November* 13, 1895, and to sell the machines so manufactured throughout the United States to the full end of the term for which said letters patent are granted.

Signed at *Brooklyn*, in the County of *Kings*, and State of *New York*, this 22*nd* day of *April*, 1896.

Witnesses: *James B. Smith.*   [Seal]
Charles Johnson,
William G. Anthony.

32. License. Not Exclusive. With Royalty.

This agreement, made this 12th day of *September*, 1896, between *James B. Smith*, of *Brooklyn*, in the County of *Kings*, and State of *New York*, party of the first part, and *Charles Johnson*, of *Clifton*, in the County of *Richmond*, and State of *New York*, party of the second part, witnesseth, that whereas letters patent of the United States No. 321,123, for an improvement in *horse rakes*, were granted to the party of the first part dated *October* 4, 1895; and whereas the party of the second part is desirous of manufacturing *horse rakes* containing said patented improvement: Now, therefore, the parties have agreed as follows:

I. The party of the first part hereby licenses and empowers the party of the second part to manufacture, subject to the conditions hereinafter named, at their factory in *Clinton*, and in no other place or places, to the end of the term for which said letters patent were granted, *horse rakes* containing the patented improvements, and to sell the same within the United States.

II. The party of the second part agrees to make full and true returns to the party of the first part, under oath, upon the first days of January and July in each year, of all horse rakes containing the patented improvements manufactured by them.

III. The party of the second part agrees to pay to the party of the first part five dollars as a license-fee upon every horse rake manufactured by said party of the second part containing the patented improvements; provided, that if the said fee be paid upon the days provided herein for semi-annual returns, or within ten days thereafter, a discount of fifty per cent. shall be made from said fee for prompt payment.

IV. Upon a failure of the party of the second part to make returns or to make payments of license-fees, as herein provided, for thirty days after the days herein named, the party of the first part may terminate this license by serving a written notice upon the party of the second; but the party of the second part shall not thereby be discharged from any liability to the party of the first part for any license fees due at the time of the service of said notice.

In witness whereof the parties above named have hereunto set their hands the day and year first above written at *Brooklyn*, in the County of *Kings*, and State of *New York*.

James B. Smith.
Charles Johnson.

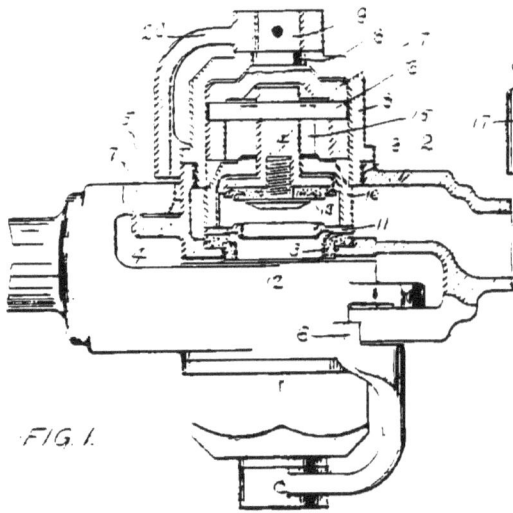

FIG. 1.

FIG 3.

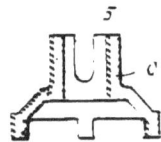

FIG 4

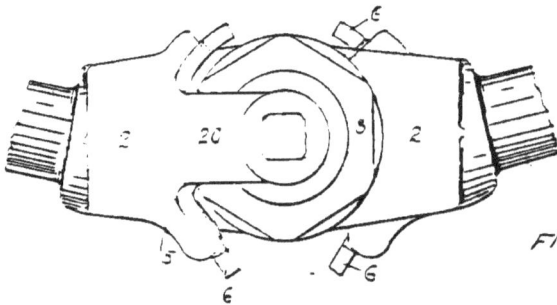

FIG 2

WITNESSES
Charles Anthony
William J. Johnson

INVENTOR
James B. Smith
by
John C. Brown, Atty

33. Complete Proceedings, Including Actions by the Office and Amendments by the Applicant.

To all whom it may concern :

Be it known that I, James B. Smith, a citizen of the United States, and resident of Brooklyn, in the County of Kings and State of New York, have invented certain new and useful improvements in Pipe Couplings, of which the following is a full and complete specification.

This invention relates particularly to pipe couplings which are attached to the air brake pipes of railroad trains ; and has for its object to do away with the stop-cock or angle-cock which controls the flow of compressed air through these air brake pipes, by the substitution of a valve in the coupling shells, to take the place of, and perform the functions of, this angle-cock. Further objects of this invention will appear later on.

In the drawings, Figure 1 is a side elevation showing two couplings in their locked position, one of such couplings being shown in section ; Figure 2 is a plan view of the same ; Figure 3 is a detail of the cam slot, and Figure 4 is a view of the chair with vertical slot.

In the construction of my improved coupling, I use a shell 2, of the type in general use, provided with a hose end 1, to which the flexible hose may be secured in any suitable manner. I provide the coupler shells with the usual gaskets 3, formed of rubber, through the aperture of which the compressed air passes. The coupler shell is also provided with the usual projecting lip 4, adapted to become engaged with the projecting arm 5, which extends outward from the body of the engaging shell. This is the ordinary and usual construction, and is so arranged that two coupler shells can be engaged with each other by a rotary motion which brings the bead 7 on the lip 4 into the groove on the arm 5 of the other couplings, which groove is provided for the reception of the bead.

To provide each coupling with a valve which shall control the opening through the gaskets, I provide a nut 8, which has a threaded portion 9 whereby the nut is firmly fitted to the shell. This nut contacts and holds in place the chair 10, which rests upon the valve seat 11. To close the opening 12, in this valve seat, I provide a valve 13, which has a reciprocating motion directly towards and away from the seat 11, for the purpose of opening and closing the aperture. This valve has a cylindrical stem 14, guided in the neck of the chair 10 ; this neck is also provided with a slot 15, in which rides the pin 16 which is also attached to the valve stem 14. Fitted within the nut 8 is the cylindrical cam 17, whose neck 18 has a bearing in the nut, and to a squared projection 19 of this neck, is fitted the lever 20. The cam 17 has a slot 21, which carries the pin 16, and by means of which the pin 16, and with it the valve 13, is moved towards and away from the seat 11. To operate the lever 20, I provide it with two projecting lugs 6 which are adapted to straddle the arm 5 of the engaging coupling. As the couplings are rotated to their engaged position, this lever is operated to rotate the cam and thus raise the valve from its seat, leaving a free opening through the gaskets. As the couplings are rotated to their disengaged position the valve is moved back to its seat, thus closing the opening.

Thus it will be seen that the operation of bringing the couplings into engagement opens the valve while that of uncoupling the shells closes the valve. Those familiar with the art will remember that in the ordinary and usual arrangement, when the couplings are engaged the angle-cock has to be turned independently, to allow the air a free passage through the train pipe, and when the shells are uncoupled the angle-cock has to be closed. Another advantage of this invention is that by bringing the valve close to the mouth of the coupling, the latter does not so readily become a lodging place for dirt and cinders, which are thence blown into the triple valve.

The features of this invention which I believe to be novel, consist in the rotary cam and in the chair by which it is supported. By this construction the valve moves to and from its seat with a direct motion, in distinction from a previous form of construction in which the valve is moved from its seat by a twisting motion which has a bad effect upon the valve, causing it to quickly wear out, and in which form of construction, it is sometimes hard to open the valve.

Having described my invention I claim :

Rejected.
1. In pipe couplings, the combination of two engaging coupler shells, a reciprocating valve within each shell, a rotary cam within each shell adapted to operate said valve as said cam is rotated, a lever attached to each cam and adapted to be operated by the opposite coupler shell as the shells are engaged and disengaged respectively, as and for the purpose set forth.

Rejected.
2. In pipe couplings, the combination of a shell, a reciprocating valve within the shell, a rotary cam connected with such valve, and a lever attached to said cam and adapted to rotate said cam, as and for the purpose set forth.

Rejected.
3. In pipe couplings, the combination of a shell provided with a valve seat, a reciprocating valve mounted upon a stem and connected with a cam,

said cam having an operating lever attached to it, as and for the purpose set forth.

4. In pipe couplings, the combination of a shell, a valve seat within said shell, a rotary cam, a reciprocating valve provided with a stem, a cross piece upon such stem and contacting said cam, a neck to said cam and an operating lever attached to said neck, as and for the purpose set forth.

5. In pipe couplings, the combination of two engaging coupler shells, a rotary cam within each shell, a reciprocating valve provided with a stem portion, a cross piece upon said stem and working in a slot in said cam, a valve seat within each shell, a chair to hold such seat in place and provided with a slotted neck in which said cross piece works, and which guides said valve stem, a lever attached to said cam and engaging the opposite coupler shell, as and for the purpose set forth.

JAMES B. SMITH.

Witnesses,
CHARLES JOHNSON,
WILLIAM G. ANTHONY.

UNITED STATES PATENT OFFICE.

WASHINGTON, D. C., Feb. 9, 1896.

JAMES B. SMITH,   Subject : Pipe Couplings.
Care John C. Brown,   Filed Jan. 2, 1896.
Brooklyn, N. Y.   Serial No. 534,743.

Please find below a communication from the Examiner in charge of your application, above noted.

JOHN S. SEYMOUR,
Commissioner of Patents.

This case has been examined.

Claims 1, 2, 3 and 4 are rejected on West, No. 214,336, Apr. 15, 1879, Water Distribution, Pipe Couplings, Valved, or on the British patent to West, No. 5201, Dec. 18, 1878, Pipe Couplings, in view of Farrell, No. 490,227, Jan. 17, 1893, Water Distribution, Cocks and Faucets, Reciprocating Valves. There would be no invention in substituting the valve shown in Farrell for that shown in either of the other patents cited.

Claim 5 may be allowed.

Explanation of the Rejected Claims.

The two patents to West show a shell similar to that of the applicant, having a reciprocating valve operated by a cam. While the cam is stationary, the valve and its stem are turned by a connection with a lever similar to the applicant's. The patent to Farrell shows a reciprocating valve operated by a rotary cylinder which makes a complete turn to raise the valve from its seat, and the slot by which this is accomplished has a regular pitch similar to a screw thread. Evidently in view of these prior patents the only patentable features of this application are the details of construction, and the four claims rejected are justly so. The claim allowed is a very narrow one and could be easily avoided by changing some of the details. To get the best protection possible under the circumstances it becomes necessary to get as many concise and distinct claims on the details as possible. Therefore the following amendment is made.

James B. Smith,
Pipe Couplings.
Filed Jan. 2, 1896.
Serial No. 534,743.

BROOKLYN, N. Y., Feb. 15, 1897.

COMMISSIONER OF PATENTS,
Washington, D. C.

Dear Sir:—In the matter of the above mentioned application and with reference to the Office letter of the 9th inst., I desire to amend as follows:

Page 2, of the specification, immediately preceding the words "Having described my invention, I claim," insert the following paragraph: As the reciprocating valve, when closed, is held to its seat by the pressure of the confined air, which is about 60 pounds to the square inch, it requires the application of considerable strength to lift it.

Therefore the slot 21 of the cam is constructed of such a form that the couplings may be rotated towards their engaged position sufficiently for the bead 7 to have entered the groove in the arm 5 and for the couplings to become partially engaged before the rise in the cam slot begins to raise the pin 16 and the valve from its seat. This rise is made very gradual at first and then more steep because the valve moves hardest at the start, and after once started it moves more easily.

As this subject matter was not described in the original specifition, and as I desire to make suitable claims thereon, I desire that the appended oath, as required in such cases, may be attached.

I desire to cancel claims 1, 2, 3 and 4, and substitute the following claims :

Allowed. 1. In a pipe coupling, the combination of a shell, a reciprocating valve within such shell, a valve seat, and a chair adapted to hold such seat in place, and to directly prevent the rotation of such valve upon its seat, and a lever operated cam adapted to control such valve, as and for the purpose set forth.

Rejected. 2. In a pipe coupling, the combination of a cam provided with a bearing in such coupling, a reciprocating valve operated by such cam, a valve seat and a means of directly preventing the rotation of said valve upon its seat, and a lever connected with such cam, as and for the purpose set forth.

Allowed. 3. In a pipe coupling, the combination of a shell, a reciprocating valve, a pin directly connected with such valve, a cam connected with such pin, and a lever attached to said cam, as and for the purpose set forth.

Allowed. 4. In pipe couplings, the combination of two engaging coupling shells, a reciprocating valve within such shell, a valve seat and a chair to hold such seat in place, said chair adapted to guide said valve and to prevent its rotation during its reciprocating motion, a cam connected with such valve, and a lever connected with said cam and adapted to be operated by the opposite shell, as and for the purpose set forth.

Allowed. 5. In a pipe coupling, the combination of a shell, a reciprocating valve, a lever operated cam connected with said valve and provided with a flat portion on its operating surface whereby said cam may be partially rotated in its opening movement without operating such valve, as and for the purpose set forth.

In order to make more plain the advantages of this construction, and the patentability of the claims submitted in the amendment I wish to ask the examiners' attention to the following :

The claims rejected in Office letter of Feb. 9th, 1896, have been cancelled, not because the applicant admits that they fully cover the structures cited against them, but in order to embody said claims in a form which will more plainly present the patentable features of this invention. It is noted that the cases cited represent two different arts. The Farrell valve is adapted to water distribution, and not to air, while the West valves are in the same art as the applicant's structure. It would therefore seem that the applicant ought to be allowed broader claims than if the Farrell valve was in the same art, since the details of construction in the two arts can not from the nature of the case be the same, and what will be operative in one art will not be an operative construction in the other.

The conditions presented to the applicant and difficulties to be overcome were these: Given a coupling shell of a size and form limited, in order to be interchangeable with those now in use, an opening in the shell of a certain size, for the same reason, to provide a valve which will control such opening and which will operate readily under a pressure of 60 pounds, and which will be operated by the action of engaging and disengaging the shells. A reciprocating valve is provided. This is shown in the West patents cited. In order that the couplings may be partially engaged before the valve is moved from its seat the cam must have flat place so that it may be turned a portion of its movement before the rise in the cam meets the pin and begins to raise the valve. By this construction the valve has no movement except the reciprocating motion directly to and from the seat. The West valve moves up with a twisting motion like that of a screw, and on moving to and from its seat must grind more or less upon it, and moreover may become wedged upon it. Hence may be seen the advantages of the applicant's construction over all previous constructions, which would seem to entitle him to claims sufficiently broad to protect his invention, and it would seem that the claims here presented ought to be allowed.

Respectfully yours,

JOHN C. BROWN,

Attorney for Applicant.

OATH.

State of New York, } ss.
County of Kings. }

James B. Smith, whose application for letters patent for an improvement in pipe couplings (serial number 534,743), was filed in the United States Patent Office on the 2nd day of January, 1896, being duly sworn, deposes and says that he verily believes himself to be the original and first inventor of the improvement as described and claimed in the foregoing amendment, in addition to that which was embraced in the claims originally made, and he does not know and does not believe that the same was ever before known or used, and that the matter sought to be inserted formed a part of his original invention at the date of filing such application, and was invented by him before he filed the same.

JAMES B. SMITH.

Subscribed and sworn to before me this 15th day of February, 1896.

ANTHONY JONES,
[Seal]   Notary Public.

UNITED STATES PATENT OFFICE.
WASHINGTON, D. C., March 18, 1896.

JAMES B. SMITH,   Subject : Pipe Couplings.
Care John C. Brown,   Filed Jan. 2, 1896.
Brooklyn, N. Y.   Serial No. 534,743.

Please find below a communication from the examiner in charge of your application.

This case has been examined as amended.

Claim 2 is rejected on the references cited, and as being functional. The remaining claims are allowable.

EXPLANATION.

Reference to the definition of Functional claims on page 13, will show that this claim falls under the class No. 3, in that the "means of directly preventing the rotation of said valve upon its seat" involves a novel element of the construction, and hence is not allowable.

James B. Smith,
Pipe Couplings.
Filed Jan. 2, 1896.
Serial No. 534,743.

BROOKLYN, N. Y., March 25, 1896.
COMMISSIONER OF PATENTS :
Washington, D. C.

Dear Sir :—In the matter of the above mentioned application, and with reference to Office letter of the 18th inst., I desire to amend as follows :

By cancelling claim 2.

In view of this action I would respectfully ask for an early allowance of this case.

JOHN C. BROWN,
Attorney for Applicant.

Note : The allowance was then granted.